SCIENTIFIC AMERICAN | EDUCATIONAL PUBLISHING

SCIENTIFIC AMERICAN INVESTIGATES
DINOSAURS AND OTHER PREHISTORIC REPTILES

ANKYLOSAURUS

BY P. J. FREEDMAN

Published in 2026 by The Rosen Publishing Group
in association with Scientific American Educational Publishing
2544 Clinton Street, Buffalo NY 14224

Portions of this work were originally authored by Leigh Rockwood and published as *Ankylosaurus*. All new material in this edition is authored by P. J. Freedman.

Library of Congress Cataloging-in-Publication Data
Names: Freedman, P. J. author
Title: Ankylosaurus / P.J. Freedman.
Description: Buffalo, NY : Scientific American Educational Publishing, an imprint of Rosen Publishing, 2026. | Series: Scientific American investigates dinosaurs and other prehistoric reptiles | Includes index.
Identifiers: LCCN 2025019487 (print) | LCCN 2025019488 (ebook) | ISBN 9781538313404 (library binding) | ISBN 9781538313398 (paperback) | ISBN 9781538313411 (ebook)
Subjects: LCSH: Ankylosaurus–Juvenile literature
Classification: LCC QE862.O65 F735 2026 (print) | LCC QE862.O65 (ebook) | DDC 567.915–dc23/eng/20250710
LC record available at https://lccn.loc.gov/2025019487
LC ebook record available at https://lccn.loc.gov/2025019488

Designer: Tanya Dellaccio Keeney
Editor: Therese Shea

Photo credits: Cover Shutterstock AI Generator/Shutterstock.com; pp. 5, 7 (top), 11 Orla/Shutterstock.com; p. 7 (bottom) Noiel/Shutterstock.com; p. 9 AKKHARAT JARUSILAWONG/Shutterstock.com; p. 13 (top) https://upload.wikimedia.org/wikipedia/commons/9/9c/Ankylosaurus_tooth.jpg; p. 13 (bottom) 3dMediSphere/Shutterstock.com; p. 15 Warpaint/Shutterstock.com; p. 17 https://upload.wikimedia.org/wikipedia/commons/4/49/Cal_Orcko_fossil_tracks_3.jpg; p. 19 paleontologist natural/Shutterstock.com; p. 21 (left) Innakote/Shutterstock.com; p. 21 (right) TimeLineArtist/Shutterstock.com.

Printed in China

CPSIA compliance information: Batch #QWSA26. For Further Information contact Rosen Publishing at 1-800-237-9932.

CONTENTS

AMAZING *ANKYLOSAURUS* 4

BIG BODY 6

ARMOR 8

TERRIBLE TAIL 10

PLANT FOOD 12

TOUGH PREY 14

WHERE DID THEY LIVE? 16

FINDING *ANKYLOSAURUS* 18

THE LATE CRETACEOUS 20

GLOSSARY 22

FOR MORE INFORMATION 23

INDEX 24

Words in the glossary appear in **bold** type the first time they are used in the text.

AMAZING *ANKYLOSAURUS*

Ankylosaurus was a big dinosaur! It was the largest of the **armored** dinosaur group known as ankylosaurids, or ankylosaurs. There were plenty of other dinosaurs that were larger than it, though.

Ankylosaurus had **adaptations** that allowed it to stay safe, even though it lived at the same time as meat-eating predators like *Tyrannosaurus rex*. Because of these adaptations, Ankylosaurus was not an easy meal—so it was eaten less often!

This picture shows how *Ankylosaurus* may have looked when it walked Earth.

FUN FACT

LIKE ALL DINOSAURS, *ANKYLOSAURUS* WAS A REPTILE. REPTILES ARE COLD-BLOODED, HAVE A BACKBONE, AND ARE COVERED IN SCALES.

BIG BODY

Ankylosaurus was not very tall, but it was long and wide. When fully grown, this dinosaur was about 25 to 35 feet (7.6 to 11 m) long. That's as long as a school bus! It was about 6 feet (2 m) wide. *Ankylosaurus* stood about 4 feet (1 m) tall and weighed 4 tons (3.6 mt) or more!

Ankylosaurus had short legs compared to the rest of its body. Its back legs were longer than its front legs.

FUN FACT

ANKYLOSAURS LIKE *ANKYLOSAURUS* WERE NOT AN ABUNDANT GROUP OF DINOSAURS. THAT MEANS THERE WEREN'T VERY MANY OF THEM.

Compared to an adult man today, *Ankylosaurus* wasn't very tall. But it was big in other ways!

ARMOR

Ankylosaurus is best known for the thick plates that covered its body. The bony plates are called osteoderms. This "armor" kept its body safe. An *Ankylosaurus*'s underside did not have any plates though.

Ankylosaurus had more than just bony plates. Two rows of spikes ran the length of its body. It also had horns on its head. It is thought that the horns did not point out toward predators. Instead, some **paleontologists** think the horns pointed up and helped males draw in females.

Ankylosaurus plates are called armor because they work a bit like the metal suits—armor—knights wore long ago to stay safe in battle.

FUN FACT

THE OSTEODERMS, ALSO CALLED SCUTES, WERE DIFFERENT SHAPES AND SIZES AND COVERED MOST OF AN *ANKYLOSAURUS* BODY.

TERRIBLE TAIL

Ankylosaurus kept itself safe with another adaptation. It had a clublike tail, which it could use to hit predators that got too close.

The vertebrae, or tailbones, of the tail were joined together. The tail was likely very stiff and hard for the dinosaur to move. It's likely *Ankylosaurus* could swing it only from side to side. If *Ankylosaurus* hit an enemy with its tail, though, that unlucky dinosaur would have felt some pain!

FUN FACT

THE *ANKYLOSAURUS'S* BRAIN WAS SMALL COMPARED TO THE SIZE OF ITS BODY. BECAUSE OF THIS, PALEONTOLOGISTS THINK THIS DINOSAUR WAS NOT VERY SMART.

At the very end of *Ankylosaurus*'s tail were more osteoderms like those on its sides and back.

PLANT FOOD

Ankylosaurus had a beaklike mouth. It had small teeth that it used to **graze** on low-lying plants, grasses, and bushes. The *Ankylosaurus* had to eat a large number of plants to keep its big body going.

Paleontologists think that this dinosaur had special organs to help it break down the large amount of plant matter it ate. Such organs would have made the plants **ferment** in the dinosaur's gut. This means that the *Ankylosaurus* would have passed a lot of gas!

ANKYLOSAURUS TEETH

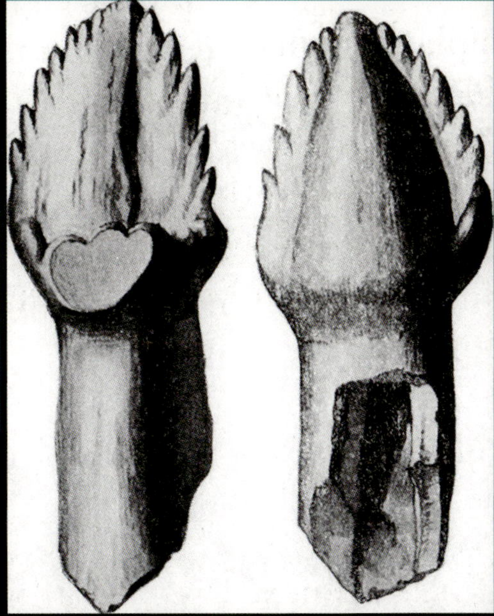

Ankylosaurus teeth looked a lot like those of the *Stegosaurus*, another plant-eating dinosaur that lived about 150 million years ago.

STEGOSAURUS

FUN FACT

THE *ANKYLOSAURUS* WAS AN HERBIVORE. AN HERBIVORE IS AN ANIMAL THAT EATS ONLY PLANTS.

TOUGH PREY

Can you imagine a huge *Ankylosaurus* moving toward you with its club tail ready to strike? Many predators, faced with the same sight, likely looked elsewhere for food.

It is thought that *Ankylosaurus* was not hunted as often as other dinosaurs of its time and place were. The main reason for this is that *Ankylosaurus* would have been harder to kill than other dinosaurs. Unless a predator could flip an *Ankylosaurus* over, its bony armor and powerful tail kept it safe.

Plant-eating dinosaurs like the *Ankylosaurus* were prey for meat-eating dinosaurs like the *Tyrannosaurus rex*. Other predators included *Deinonychus* and *Tarbosaurus*.

FUN FACT

***ANKYLOSAURUS*'S SHORT LEGS MADE IT HARD TO FLIP OVER—ANOTHER ADAPTATION THAT STOPPED PREDATORS!**

WHERE DID THEY LIVE?

Because *Ankylosaurus* needed so much plant matter to eat, it lived in warm, wet places where lots of plants grew. *Ankylosaurus* lived on what we now call the **continent** of North America.

Ankylosaurus **fossils** have been found in Montana and in Canada. At the time, these lands were warm and full of plant life. Today these places are mostly colder and drier than they were when *Ankylosaurus* lived. **Fossilized** ankylosaur tracks have also been found in the country of Bolivia, in South America.

FUN FACT

IT'S LIKELY *ANKYLOSAURUS* WAS A SLOW-MOVING DINOSAUR MOST OF THE TIME.

Many kinds of dinosaur tracks have been found at Cal Orck'o in Bolivia, including those made by ankylosaurs.

FINDING ANKYLOSAURUS

The tracks found in Bolivia have helped scientists learn how *Ankylosaurus* likely moved around. First, they looked at the distance between the tracks and the way in which the dinosaur's feet fell. Then, they compared this with the length of the legs on an *Ankylosaurus* fossil.

Finding other *Ankylosaurus* fossils has allowed paleontologists to study this amazing dinosaur. The first *Ankylosaurus* fossils were found in the early 1900s. However, very few *Ankylosaurus* fossils have been found. A complete skeleton has yet to be uncovered.

By studying fossils, scientists have come up with new theories, or ideas, about animals that have been extinct for millions of years.

FUN FACT

THE PALEONTOLOGIST BARNUM BROWN GAVE *ANKYLOSAURUS* ITS NAME IN 1908.

THE LATE CRETACEOUS

Ankylosaurus lived from about 68 to 66 million years ago. On the geologic time scale, this was during the Late Cretaceous period. At the end of this period, dinosaurs became extinct, or died out completely.

Paleontologists have many ideas about why this mass extinction happened. Some think it was caused by **climate change** or **volcanic** activity. Based on the fossils that have been found, paleontologists believe that *Ankylosaurus* was one of the last dinosaurs to become extinct.

Scientists divide the long history of Earth up into parts. This is called the geologic time scale. You can see here that the Cretaceous period is part of the Mesozoic era.

FUN FACT

THE WORD "LATE" TELLS US THAT *ANKYLOSAURUS* LIVED AT THE VERY END OF THE CRETACEOUS PERIOD.

GEOLOGIC TIME SCALE

Eon	Era	Period		Epoch	Millions of Years Ago
Phanerozoic	Cenozoic	Quaternary		Holocene	0.0117
				Pleistocene	2.6
		Tertiary	Neogene	Pliocene	5.3
				Miocene	23
			Paleogene	Oligocene	34
				Eocene	56
				Paleocene	66
	Mesozoic	Cretaceous			145
		Jurassic			201
		Triassic			252
	Paleozoic	Permian			298
		Carboniferous	Pennsylvanian		323
			Mississippian		359
		Devonian			419
		Silurian			443
		Ordovician			485
		Cambrian			541
Precambrian	Proterozoic eon				2500
	Archean eon				4000
	Hadean eon				4600

GLOSSARY

adaptation: A change in a type of animal that makes it better able to live in its surroundings.

armored: Having armor, which is the outer covering of an animal that keeps its body safe from harm.

climate change: Long-term change in Earth's climate, caused mainly by human activities such as burning oil and natural gas.

continent: One of Earth's seven large landmasses.

ferment: To change in a way that makes gas bubbles.

fossil: The hardened remains of dead animals or plants.

fossilize: To change into a fossil.

graze: To feed on plant matter.

paleontologist: A scientist who studies the past using fossils.

volcanic: Having to do with a volcano, which is an opening in a planet's surface through which hot, liquid rock sometimes flows.

FOR MORE INFORMATION

Books

Jackson, Tom. *T. Rex vs. Ankylosaurus: Prehistoric Showdown.* Minneapolis, MN: Lerner Publications, 2026.

Raatma, Lucia. *Ankylosaurus.* Ann Arbor, MI: Cherry Lake Publishing, 2025.

Websites

Ankylosaurus
https://kids.nationalgeographic.com/animals/prehistoric/facts/ankylosaurus
Read all about *Ankylosaurus* and other dinosaurs on the National Geographic Kids website.

Ankylosaurus
https://www.nhm.ac.uk/discover/dino-directory/ankylosaurus.html
Check out more about *Ankylosaurus* on the UK's Natural History Museum website.

Publisher's note to educators and parents: Our editors have carefully reviewed these websites to ensure that they are suitable for students. Many websites change frequently, however, and we cannot guarantee that a site's future contents will continue to meet our high standards of quality and educational value. Be advised that students should be closely supervised whenever they access the internet.

INDEX

adaptations, 4, 10

armor, 4, 8, 14

Bolivia, 16, 17, 18

brain, 10

Brown, Barnum, 19

Deinonychus, 15

diet, 12, 13, 15, 16

fossils, 16, 18, 19, 20

geologic time scale, 20, 21

Late Cretaceous period, 20, 21

mass extinction, 20

North America, 16

osteoderms, 8, 9, 11

size, 4, 6, 7

Stegosaurus, 13

tail, 10, 14

Tarbosaurus, 15

teeth, 12, 13

Tyrannosaurus rex, 4, 15